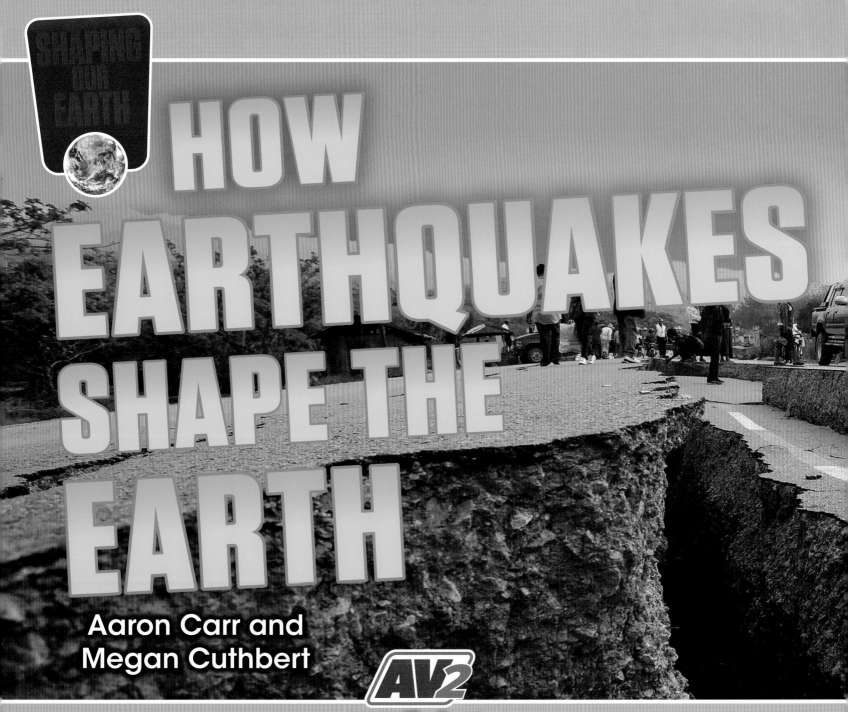

SHAPING OUR EARTH

HOW EARTHQUAKES SHAPE THE EARTH

Aaron Carr and Megan Cuthbert

AV2

www.av2books.com

Step 1
Go to **www.av2books.com**

Step 2
Enter this unique code

QUWTXOULG

Step 3
Explore your interactive eBook!

HOW EARTHQUAKES SHAPE THE EARTH

SHAPING OUR EARTH

Start!

Your interactive eBook comes with...

AV2 is optimized for use on any device

Read

Audio
Listen to the entire
book read aloud

Videos
Watch informative
video clips

Weblinks
Gain additional
information for research

Try This!
Complete activities and
hands-on experiments

Key Words
Study vocabulary, and
complete a matching
word activity

Quizzes
Test your knowledge

Slideshows
View images and captions

View new titles and product videos at
www.av2books.com

HOW EARTHQUAKES SHAPE THE EARTH

Contents

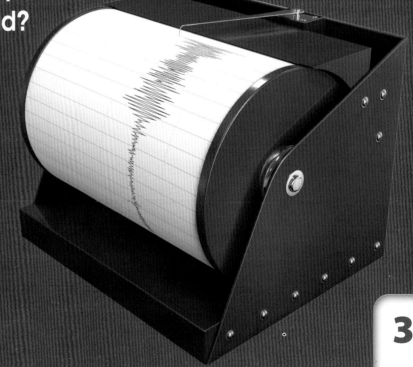

How Do Earthquakes Shape the Earth?

The Earth is always changing. Some changes happen very quickly. Many of the most Earth-shattering changes are made by earthquakes. Earthquakes can break apart the ground. People can feel the ground moving underneath them. This can be very scary.

5

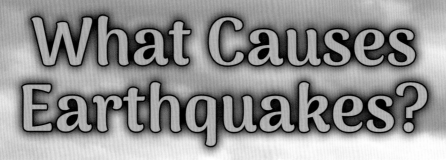

What Causes Earthquakes?

Land sits on top of large layers of rock deep under the ground. These rock layers are like big plates. The plates are always moving. Sometimes they crash into each other. This causes an earthquake.

More Earthquake Causes

Some earthquakes are caused by volcanoes. A volcano can shake the Earth when it shoots out lava and smoke. People can also cause earthquakes. This may happen when people dig deep underground to gather coal or oil.

Where Do Earthquakes Happen?

Most earthquakes happen where plates meet under the ground. Cracks often form along the edges of these plates. These cracks are called fault lines.

The **Ring of Fire** is the **largest** fault line in the world. It causes about **90 percent** of the world's earthquakes.

How Do Earthquakes Spread?

The shaking of an earthquake starts at one point. This is called the epicenter. The shaking then spreads out from the epicenter. It spreads in all directions. Very strong earthquakes can make the land shake hundreds of miles away.

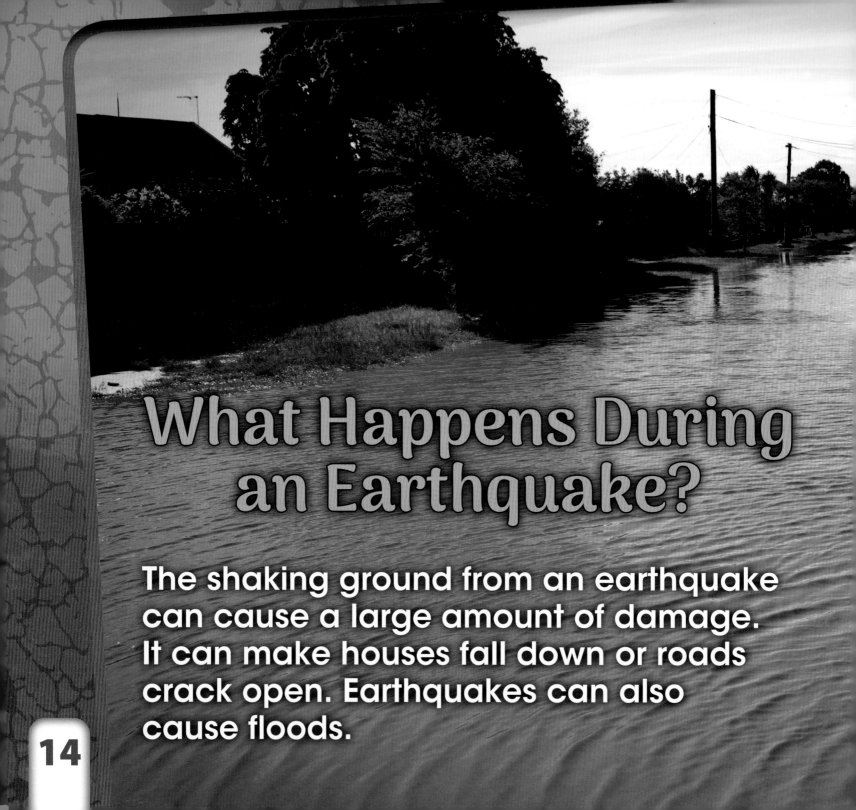

What Happens During an Earthquake?

The shaking ground from an earthquake can cause a large amount of damage. It can make houses fall down or roads crack open. Earthquakes can also cause floods.

What Changes Can Earthquakes Cause?

The strongest earthquakes can move large areas of land. Some of these strong earthquakes move land deep underwater. This pushes water up to make giant waves. These giant waves are called tsunamis.

Lituya Bay, in Alaska, was the site of the largest tsunami ever recorded. It reached up to 1,720 feet (524 meters) high.

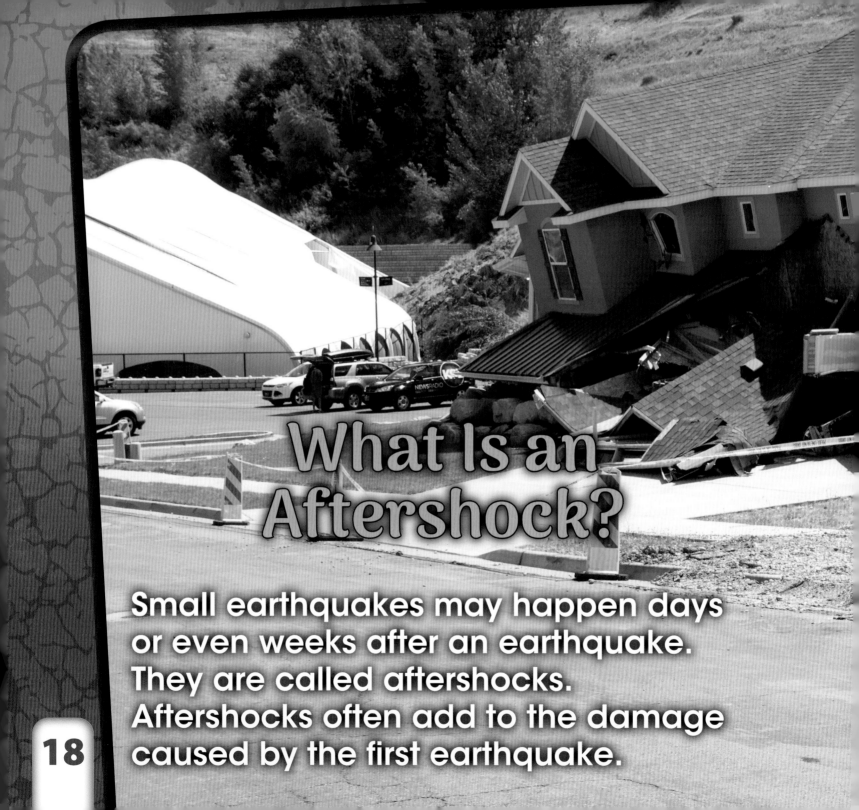

What Is an Aftershock?

Small earthquakes may happen days or even weeks after an earthquake. They are called aftershocks. Aftershocks often add to the damage caused by the first earthquake.

How Can I Stay Safe During an Earthquake?

Buildings shake during an earthquake. Experts say people should hide under a strong piece of furniture during an earthquake. It is important to stay away from falling objects. Experts also say to stay away from windows. Listen to the radio to hear what the officials say. They will know the best way to keep safe.

There are about **50 earthquakes per day** around the world.

EARTHQUAKE FACTS

These pages provide detailed information that expands on the interesting facts found in the book. These pages are intended to be used by adults to help young readers round out their knowledge of each natural event featured in the *Shaping Our Earth* series.

Pages 4–5

How Do Earthquakes Shape the Earth? Earthquakes release waves in the ground. The fastest waves move at speeds of up to 8 miles (14 kilometers) per second. Most earthquakes are small and go unnoticed by people. Large earthquakes cause more damage and have a greater impact on the surface of the Earth. Parts of the ground are ripped in different directions during an earthquake. This can cause scars on the land.

Pages 6–7

What Causes Earthquakes? The Earth has many layers. The top layer, called the crust, rests on about 15 to 20 tectonic plates. These plates fit together like a puzzle. They are constantly moving. Most earthquakes are caused by plates colliding or sliding past each other.

Pages 8–9

More Earthquake Causes The same processes that create volcanoes can cause earthquakes. Pressure builds beneath the tectonic plates from the upward movement of magma, a liquefied rock substance. When the magma bursts through cracks in the plates, it causes a small earthquake. The drilling that oil companies use to gather oil has also been shown to cause earthquakes.

Pages 10–11

Where Do Earthquakes Happen? Earthquakes occur most often along the edges of tectonic plates. When tectonic plates are moving in opposite directions, cracks in the Earth, called faults, are created. These faults are usually located near the edges of the plates.

How Do Earthquakes Spread? Earthquakes begin at cracks in the tectonic plates. An earthquake's origin point is called its focus. The epicenter is directly above the focus. The crack at the focus sends waves of movement, called seismic waves, that can be felt far away from the epicenter. Scientists use a scale called the Richter scale to measure earthquake strength.

What Happens During an Earthquake? Earthquakes can change the land dramatically. Earthquake vibrations can trigger landslides. Land can also be raised or dropped. A 2010 Chilean earthquake raised the ground by 8 feet (2.5 m) in some areas and lowered the ground in others. The earthquake also moved the city of Concepción 10 feet (3 m) to the west.

What Changes Can Earthquakes Cause? Powerful earthquakes can move large amounts of land. An Indonesian earthquake in 2004 moved the location of the North Pole almost 1 inch (2.5 centimeters) to the east and slightly changed the shape of the Earth. The movement of the land deep underwater can cause giant waves, called tsunamis. These tsunamis build in the ocean and travel great distances. When they reach land, tsunamis can flood the land and cause immense devastation.

What Is an Aftershock? Additional tremors after an earthquake are called aftershocks. Most occur within one fault-length's distance from the epicenter. Aftershocks happen as the displaced rock settles into its new position. They can do substantial damage. In 2011, an aftershock in Christchurch, New Zealand, was deadlier than the earthquake that preceded it a few months earlier.

How Can I Stay Safe During an Earthquake? Much of the danger during earthquakes comes from falling debris and collapsing buildings. If a person is indoors, load-bearing walls or sturdy furniture can provide protection. If a person is outside, he or she should stay far from trees, power lines, and buildings.

KEY WORDS

Research has shown that as much as 65 percent of all written material published in English is made up of 300 words. These 300 words cannot be taught using pictures or learned by sounding them out. They must be recognized by sight. This book contains 86 common sight words to help young readers improve their reading fluency and comprehension. This book also teaches young readers several important content words, such as proper nouns. These words are paired with pictures to aid in learning and improve understanding.

Page	Sight Words First Appearance	Page	Content Words First Appearance
4	always, are, be, by, can, changes, do, Earth, how, is, made, many, most, of, people, some, the, them, this, very	4	earthquakes, ground
7	an, big, each, into, land, large, like, on, other, sometimes, these, they, under, what	7	layers, plates, rock
8	a, also, and, it, may, more, or, out, to, when	8	coal, lava, oil, smoke, volcanoes
10	about, in, line, world	11	cracks, edges, fault lines
11	along, often, where	13	directions, epicenter, point
13	all, at, away, from, make, miles, one, point, starts, then	14	amount, damage, floods, roads
14	down, great, open	16	areas, tsunamis, waves
16	move, up	18	aftershocks, weeks
17	feet, high, was	20	buildings, experts, furniture, objects, officials, piece, radio, windows
18	add, after, days, even, first, small		
20	hear, I, important, keep, know, say, should, way, will		
21	around, there		

Published by AV2
350 5th Avenue, 59th Floor New York, NY 10118
Website: www.av2books.com

QE534.3 .C37 2021 (print) | LCC QE534.3 (ebook) | DDC   551.22--dc23 LC record available at https://lccn.loc.gov/2020005330 LC ebook record available at https://lccn.loc.gov/2020005331

Printed in Guangzhou, China
1 2 3 4 5 6 7 8 9 0 24 23 22 21 20

042020
100919

Project Coordinator: Sara Cucini   Designer: Ana María Vidal

AV2 acknowledges Getty Images and Shutterstock as the primary image suppliers for this title.

Library of Congress Cataloging-in-Publication Data

Names: Carr, Aaron, author. | Cuthbert, Megan, author. Title: How earthquakes shape the earth / Aaron Carr and Megan Cuthbert. Description: New York, NY : AV2, [2021] | Series: Shaping our earth | Audience: Grades 2-3 | Identifiers: LCCN 2020005330 (print) | LCCN 2020005331 (ebook) | ISBN   9781791125585 (library binding) | ISBN 9781791125592 (paperback) | ISBN   9781791125608 (ebook other) | ISBN 9781791125615 (ebook other) Subjects: LCSH: Earthquakes--Juvenile literature. | Earth sciences--Juvenile literature. Classification: LCC